Okoye Uchechukwu

Field Development Programme for Holly Field

GRIN Publishing

Bibliographic information published by the German National Library:

The German National Library lists this publication in the National Bibliography; detailed bibliographic data are available on the Internet at http://dnb.dnb.de .

Imprint:

Print and binding: Books on Demand GmbH, Norderstedt Germany
ISBN: 978-3-656-97856-5

This book at GRIN:

http://www.grin.com/en/e-book/300250/field-development-programme-for-holly-field

School of Engineering

Due Date	Date Submitted	For official use only
30/11/2014		LATE DATE

SURNAME	OKOYE
FIRST NAME(S)	UCHECHUKWU
COURSE & STAGE	MSc Oil & Gas Engineering
ASSIGNMENT TITLE	Field Development Programme for Holly Field

Table of Contents

List of Figures

List of Tables

Executive summary

This report presents the Holly field exploratory well development programme including well planning and designing which requires engineering support for optimum drilling operations. The exploratory well to be drilled lies 12,000ft (true vertical depth) relative to the ground.

The report is based on the discussion of the field development plan options considered which includes the functions of the three rotary systems (hoisting, circulating and rotating) which are required in the rotary drilling. Also, it explains the main criterion for selecting a land rig and how the accurate values of pore and fracture pressures are important for successful safe drilling of the well. The function of the additives used in Portland cement for casing cementing were stated; cementing calculation was done, determining the volume of the cement slurry and mud required to displace the cement, then differential pressure at the float collar. The methods of selecting casing setting depths were discussed. In drilling the final hole section of the well, the detail functions of various selected BHA component were stated and the calculation of the number of the drill collar required in the BHA to provide a 21,000lb weight on bit at an angle 10 degree was done. The options for completing the well and the reasons for bottom to the top design were also discussed.

1.1 Rotary drilling

Rotary drilling is any form of drilling which involves a rotary action combined with downward force to grind away the cuttings in which a hole is being formed (Azar and Robello, 2007). The three systems on a rotary drilling are the:

- Hoisting system
- Circulating system
- Rotating system

1.1.1 Hoisting system

The hoisting system on a rotary rig consist of tools used to support a means to lower and raise the principal items of equipment such as drill strings, casing strings and other necessary subsurface equipment in and out of the well. The drilling operation involves withdrawing the drill string from the well, to change a dull drill bit and to add additional drill pipe to the drill string as the hole deepens. The major components are the draw-works, block and tackle, derrick and substructure.

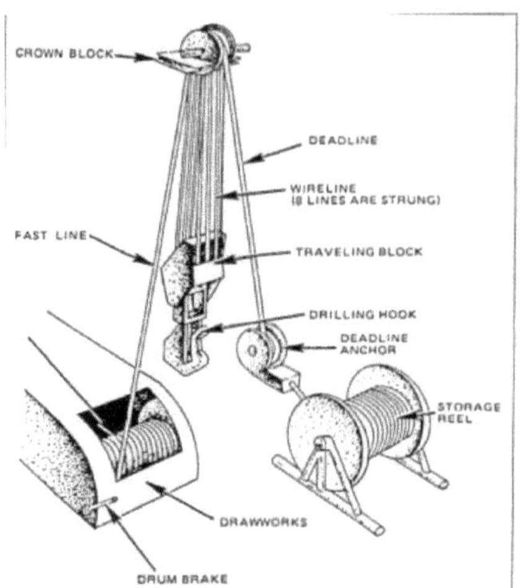

Figure 1: A rotary rig hoisting system (Mohammed, 1992)

1.1.2 Circulating system

The primary function of the fluid-circulating system in rotary drilling is to allow the movement of the mud through the well to remove the rock cuttings from the bottom hole to the surface during drilling process (Azar and Robello, 2007). It also cools and lubricates the drill bit; and in an uncased hole, it serves to support the walls of the hole. The principal components of the circulating system include mud pumps, mud pits, mud mixing and treatment equipment.

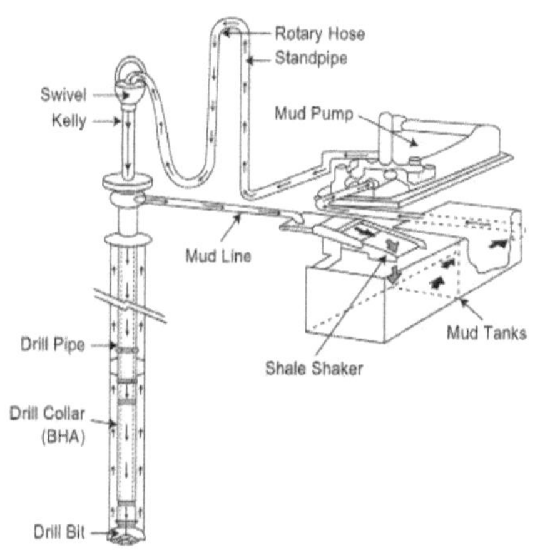

Figure 2: A mud circulating system (Boyun and Gefei, 2011).

1.1.3 Rotating system

The rotating system on a rotary drilling rig consists of the components used to achieve bit rotation. The major function involves transmitting rotating action to the drill string and consequently the bit. The main parts include drill pipe, swivel, drill collar, Kelly, rotary table and drive. (Azar and Robello, 2007).

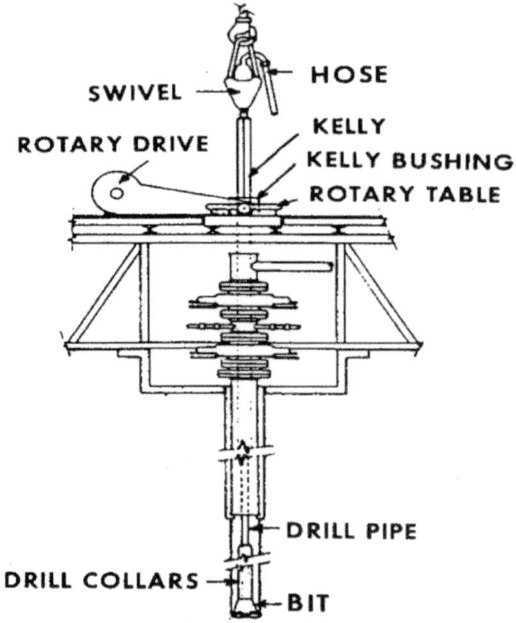

Figure 3: Schematic of rotary system (Bourgoyne et al, 1984)

1.2 Process of rotary drilling

In drilling a well for the purpose of producing oil or natural gas, there are several elements involved to successfully and economically drill a well. They include:

- Downward force on a drill bit
- The drill bit rotation
- Fluid circulation

Drilling through a rock require a force (weight of the bit) to be applied downward to the bit by the drill collars, in the drill string. The bit is rotated by drill string using the rotary table or top drive motor at the surface and the drill collar in the drilling string also applies the downward force on the bit.

During drilling process, rock cuttings and heat are being continuously generated. To remove the cuttings from the hole a drilling mud has to be continuously circulated from the surface down the drill string through the bit to the bottom, the fluid cleans the bit from the cuttings and carries the cuttings through the annular space between the drilling string and the walls of the hole to the surface again to dissipate the heat and remove cuttings and is circulated back into the drilling string. (Abdel, et al, 2003)

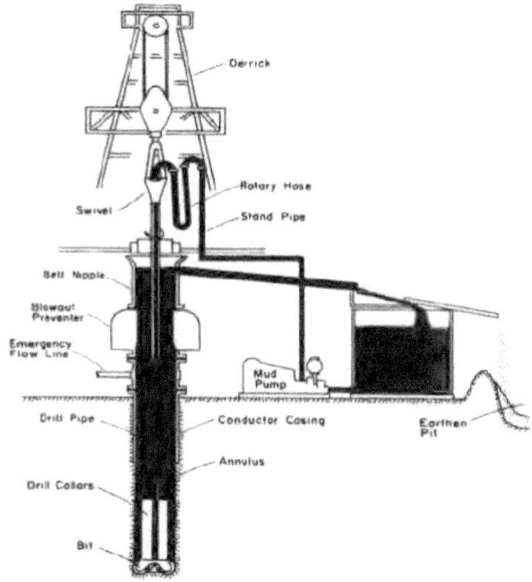

Figure 4: A rotary drilling process (Bourgoyne et al, 1984).

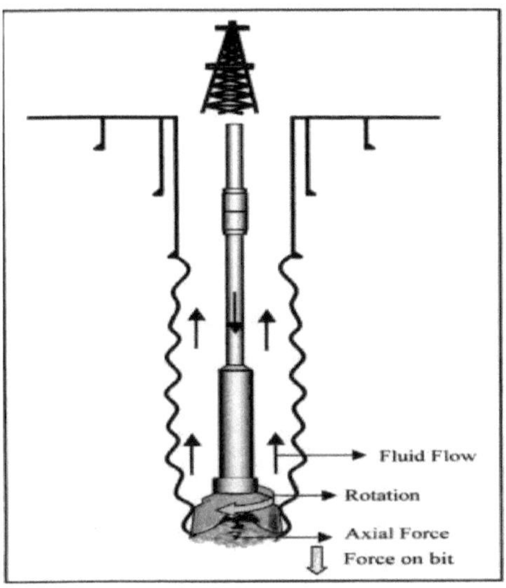

Figure 5: Basic element to drill a well (Azar and Robello, 2007).

1.3 Criterion for land rig selection

A rig is large complex machine design to drill a wellbore which comprises of most facilities for drilling purposes. Any choice of a rig is usually a compromise so in selecting a land rig for a particular well it is important to consider a well trajectory from the surface to total depth (TD) prior to drilling, providing expectations for the total depth of the drilling, lithology and pressure. The total depth of the well depends on the rig selection which should be capable of drilling the designed well with sufficient redundancy to handle all possible constraints.

2.1 Well planning

The purpose of good well planning (such as choice of mud weight profile and casing programs) and actual drilling operations is to avoid or minimise the dangers of lost circulation, well kicks, stuck pipe, blowout and excessive costs, if pressure is properly evaluated. The drilling mud density which is used to control the hydrocarbon pressure is determined by the pore and fracture pressure. Thus basic knowledge of the two key pressure parameters is important.

Pore pressure is the pressure exerted by the fluids (such as hydrocarbons) in the pore space of rock. Fracture pressure is the pressure at which the subsurface formation is not strong enough to withstand the pressure of the drilling mud in the well and is fractured.

In well planning pore and fracture pressure can help in the following:

2.1.1 Well kick

Kick occurs when the fluid from the formation flows into the wellbore during drilling operations. This can be prevented if the hydrostatic pressure of the drilling mud density is approximately equal or slightly higher than the pore pressure to prevent formation fluids from entering the wellbore (trip margin). If the mud weight is too low and the drillers do not balance the pressure, hydrocarbons can flows into the well bore causing kick; a blow-out could result if proper well control procedures are not followed. (Byrom, 2014).

2.1.2 Lost circulation

Lost circulation occurs when the drilling mud flows into the surrounding formation instead of returning to the surface. This is avoided if mud density is lower than the fracture pressure so that the drilling fluid does not fracture and enter the formation (kick margin). But if the mud weight is too high it can fracture the surrounding rock; when this happens the drilling mud flow out of the well bore into the formation instead of circulating back to the surface. This potentially causes lost return and circulation. This inflow can cause unprotected section of the well to collapse.

2.1.3 Selecting of casing setting depths

The changes in pore and fracture pressure gradient varies with depth. As the well is drilled, the pore pressure increases in the bottom of the open hole because the pressure difference between the mud and pore pressure gradient is decreasing which causes an exceeding fracture pressure of the formation higher up in this open hole. When this occurs, the drillers can no longer depend on the mud to control pore pressure in the wellbore because if they increase the mud density, it will fracture the relatively weak, exposed formation higher up and if they keep drilling without increasing the mud weight, hydrocarbons or other fluids in the deeper formation will flow into the well. Since the wellbore pressure must be maintained, then casing is set to help prevent the weak formation of the hole outside the casing from the pressure of the drilling mud inside and also protects high pressure fluids outside the casing from entering the well. Addition of casing string is necessary during drilling as the pore pressure in the formation approaches the fracture pressure at the last casing seat. This is done to prevent the mud weight from fracturing the formation at the last casing string depth. The accurate determination of pressure required to fracture a formation is important for selecting casing depth.

The complete well planning process with some exceptions, is based on prediction of pore and fracture pressure. The pressure is the key factor for many sectors of the well plan. If adequate concern is not shown to formation pressure, the other technical segment of the well plan may be affected.

3.1 Pore pressure gradient

The pore pressure gradient, expressed as an equivalent mud weight, is a curve that shows how the pore pressure in the well changes by depth.

Table 1: Pressure gradient expressed as an equivalent mud weight, (EMW) and Depth (ft)

Mud weight (ppg)	Pore pressure (EMW)	Depth (ft)
9.2		550
9.5	8.9	4000
11	10.5	9000
11	10.5	12000

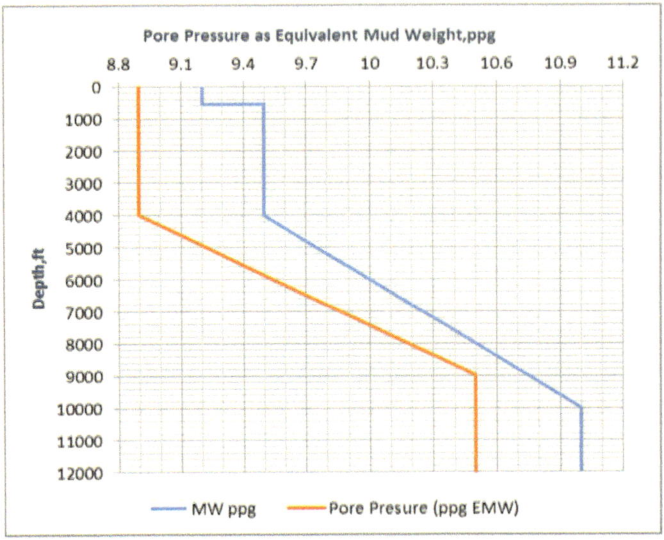

Figure 6: Pore pressure equivalent mud weight (EMW), and mud weight versus well depth.

Table 2: Determination of pore pressure at different depths.

Depth (ft)	Pore pressure (psi)
4000	1851.2
9000	4914
12000	6552

4.1 Cement additives and cementing

Cement additives have been produced to enable the use of Portland cement in many oil and gas well application. These additive modifies the cement grade to suit the well being drilled. The function and example of each of the additive includes:

4.1.1 Lightweight additives

They are known as extenders or water absorbing. It reduces cement slurry density. It reduces the hydrostatic pressure during cementing, which avoids the breakdown of weak rock formations preventing induced lost circulation caused by such breakdown. It reduces the mobility of water as more water is added to lower the weight of the slurry. It increases the viscosity of the slurry, increase the permeability and reduce the final strength of the cement. One of the example of the additive is Bentonite. (Nelson, 1990).

4.1.2 Retarders

They are used to lengthen the thickening time of oil well cement slurries or to extend the cement setting. It slows the rate of strength development. At higher temperatures it is used to allow time or placement of liquid slurry. One of the example of the additive is Calcium Lignosulfonate.

4.1.3 Accelerators

They are used to shorten the cement slurries set time or enhances the cement setting. Increases the rate of bonding strength of the cement and shorten thickening time. At low temperatures it is used to reduce time waiting on cement. They cause cement to absorb or react with water quickly. One example of the additive is calcium chloride.

4.1.4 Heavyweight additives

Used to increase oil well cement slurry densities in deep or high temperature wells. It helps restrain high formation pressures and improve mud displacement. One of the commonly used additive is Barite. (Economides, et al, 1998).

4.2 Cementing calculations

- The Volume of cement slurry required using an open hole excess of 30% = 499 bbls

Table 3: The capacity of annulus and the volume of cement

Capacity of Annulus	bbl/ft	Volume of Cement	Bbls
Capacity of 13 3/8" and 9 5/8"	0.0581	In the casing	58
Capacity of 12 ¼" and 9 5/8"	0.0558	In the open hole	435
Capacity of the shoe track	0.07076	In the shoe track	5.66
		Total	499

- The volume of mud required to displace the cement = 702.3 bbls

- Differential pressure at the float collar before the slurry starts to set = 450 psi

5.1 Selection of casing setting depths

Casing is a high strength steel pipe that comes in different foot sections that the drillers incorporate on the rig to make a casing string. Casing string help to prevent high pressure fluids outside the casing from entering the well, maintain formation stability and control well pressure. The selection of the depths at which to set the casing is very important for well design. For this purpose, the selection is performed by evaluating the subsurface data such as the pore and fracture pressure, formation stability and total depth of the well.

The key factor for selecting the number of casing strings and their respective setting depths depends on consideration of the pore and fracture pressure gradients of the well to be drilled. Selecting casing string setting depths for the given well can be difficult because of shortage of data (fracture pressure data was not given). In this situation some assumption must be made using figures 7 and 8.

The figures assumed that the behaviour of the well have been known before drilling, it shows the relationship between these gradients and casing setting depth. The pore and fracture pressure gradients data are to be obtained by the seismic and drilling data of the well, expressed as equivalent mud density, and are plotted versus depth. A broken line representing safety margins which establish the planned mud density program are introduced.

The two figures 7 and 8 shows two possible ways to estimate setting depths:

5.1.1 Bottom-top design

In this design, to reach the total depth at figure 7, the highest mud weight used to prevent the inflow of hydrocarbon into the well is required at point A. Thus, to drill with the mud weight without fracturing the shallow exposed formation within the bore hole. A vertical line is drawn upward from point A to point B where the fracture gradient equals the mud density needed to drill to point A, then a protective intermediate casing string must be set at point B, to prevent formation damage on the section AB. To select another casing depth, draw a horizontal line from the fracture gradient safety margin, point B which intersect at the pore pressure safety margin, point C, then, draw another vertical line upward from the point C to the fracture gradient safety margin point D which establish the setting point of the casing at the section DC. This similar process is continued for determination of other casing string until a specified surface casing depth is established. (Robert, 2006).

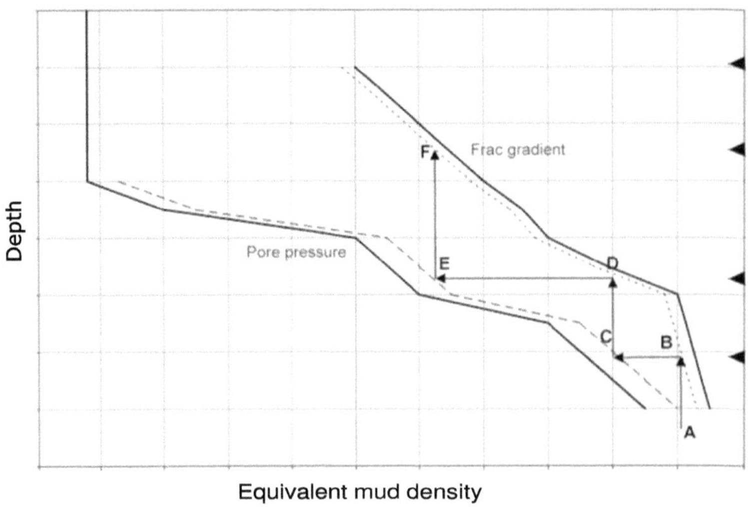

Figure 7: Casing setting depth-bottom-top design. (Robert, 2006)

5.1.2 Top-bottom design

In this design, casing depth selection is done by drawing a vertical line downwards from the fracture gradient safety margin of the surface casing, Point A to the pore pressure safety margin point B. These establishes the setting point of another casing at about point AB. Draw a horizontal line from the point B to the fracture gradient safety margin at point C, then, draw a vertical line downward to the pore pressure safety margin point. This establishes another casing setting depth. This similar process is continued until the bottom hole of the bore hole is established. (Robert, 2006).

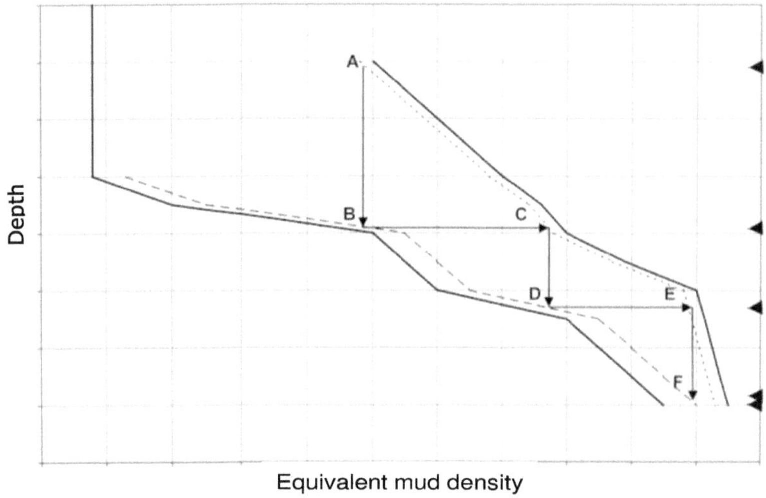

Figure 8: Casing setting depth-top-bottom design. (Robert, 2006)

6.1 Bottom Hole Assembly (BHA)

Bottom hole assembly is group of drilling component, that make up the lower end of the drill pipe. BHA is very useful in the drilling process, the adequate selection of the correct BHA will ensure high rate of penetration and obtain subsurface information. The assembly can consist of

- Drill bit (Polycrystalline diamond compact, PDC)
- Mud motor
- Measurement while drilling (MWD)
- Logging while drilling (LWD)
- Non-magnetic drill collar (NMDC)
- Drilling jar
- Drill collars
- Stabilizers
- Subs

6.1.1 Drill bit

This is the hardest drilling tool used for cutting or boring through rock strata.

6.1.2 Mud motor

The positive displacement of the downhole mud motors uses the hydraulic energy of the drilling mud to rotate the bit without the rotation of drill string. Erosion and wear are reduced on the drill string since the strings are not rotated in the process. The mud motor adds additional power to the drill bit, and also increases the Rate of penetration (ROP).

6.1.3 Measurement while drilling (MWD)

This is a type of well logging that integrates the measurement tools into the drill string and provides the real-time data such as the measurement of the direction and inclination of the wellbore, drilling tools orientation and drilling mechanic data which are transmitted up the pipe.

6.1.4 Logging while drilling (LWD)

This is a type of well logging that integrates the logging tools into the drill string and provides high quality data such as formation parameters (resistivity, porosity, sonic velocity, gamma ray) which are recorded in memory and downloaded when the tools reach the surface.

6.1.5 Non-magnetic drill collar (NMDC)

This is a non-magnetic chrome nickel steel drill collar which allows core orientation and measurement of hole direction to be taken. It also gives weight to the bits.

6.1.6 Drilling jars

They are tools used to free stuck pipe during drilling.

6.1.7 Drill collars

They are primary component of the BHA. They are the heavy steel pipe in the drill string above the bit which provides the weight and downward force needed to break rock, reduce bit stability problems and provides stiffness to the BHA to reduce directional control problem.

6.1.8 Stabilizers

They are used in the BHA of a drill string to control or minimize bending and vibration stresses, they equalizes the weight of the drill collar on the bit, they guide the bit to ensure the quality of the bore hole.

6.1.9 Subs

They are used to connect all the different parts of the BHA together. The different components of BHA usually have different connection sizes, and these subs are used as crossovers for the components. Subs can also be used as disposable components of BHA in order to prolong the life of the expensive strings they are connected to in the BHA.

6.2 Drill collars calculation

The number of drill collars required in the BHA to provide a weight on bit of 21,000 lb to drill the final section of the well is 9 Drill collars

7.0 Well completion

Well completion is the subsurface mechanical configuration of the well which provides the pathway for the hydrocarbons from the formation to the well head at the surface. (Abdel, 2003). It is an important link in oil field development because the design is needed by the engineers to plan and execute the well in the expected production target.

There are two interface options which can be further classified into different alternatives in completion architecture:

1. Upper completion (according to zones)
    - Single zone
    - Multiple zone
2. Lower completion
    - Open hole
    - Cased (Perforated) hole

The lower completion are either cased or open hole completion depending on the characterises of the hydrocarbon formations to be drilled. For the given well, the option of the lower completion is the cased (perforated) hole completion because this type of completion allow multiple reservoir production, it involves perforating the casing and cement sheath across the producing zone to establish communication between the formation and the well, the cemented casing which helps in reservoir isolation, selective inflow and gives mechanical support to the well.

The upper completion will require the multiple zone with dual zone single packer. This completion allow flow through one tubing with the help of the single packer because production is done on one reservoir and if in future the reservoir A is to be produced then the production liner of the reservoir will be perforated again for production.

7.1 Objective of bottom top design

The bottom top design provides maximum optimum production from the zone. It will ensure the well's mechanical integrity by providing a design basis for casing size and the type of casing that withstand the flow potential and the drilling problem which allow the well to be drilled and completed safely in the desired producing configuration.

8.0 Well schematic

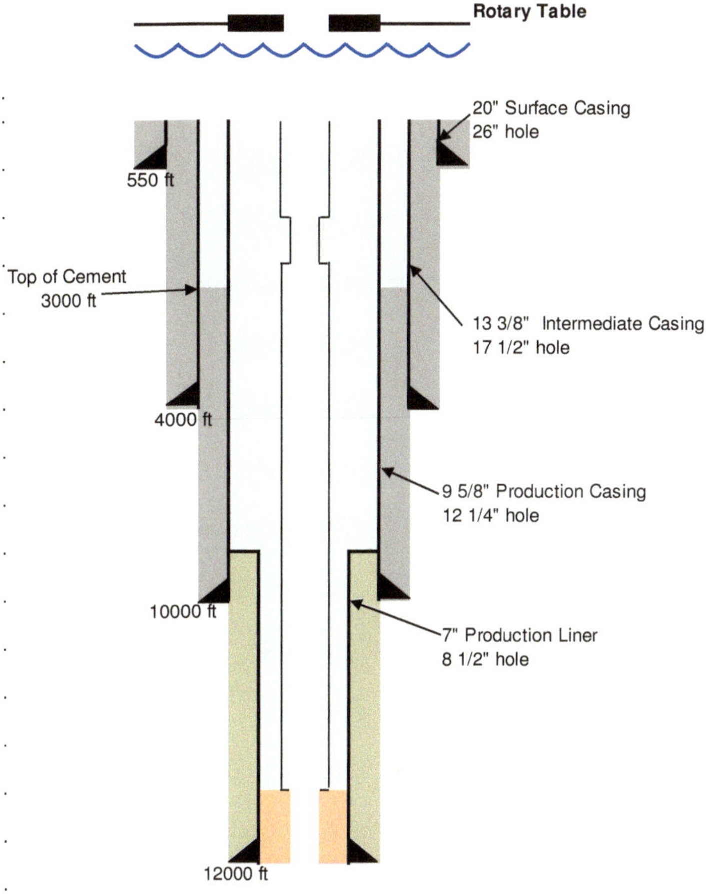

Figure 9: Schematic diagram of the well showing the casing and cementing details.

CONCLUSION

The selection of equipment, components, casing setting depths, options for well completion and calculations for field development plan of the Holly field was carried out through a careful analysis of available options for the given well and due consideration for well planning and design that will provide successful safe drilling and optimum production from the pay zone.

Words count: 2837

REFERENCES

ABDEL H.K, MOHAMMED AGGOUR, FAHIM.M.A, 2003. *Petroleum and gas processing.* Published by CRC press.

BOURGOYNE A.T, et al, 1984. *Applied Drilling Engineering.* Published by SPE publications.

BYROM TED.G, 2014.*Casing and liners for drilling and completion.* Published by Elsevier.

ECONOMIDES M.J, et al, 1998.*Petroleum well construction.* Published by John Wiley and sons.

MOHAMMED A. MIAN, 1992.*Petroleum Engineering Handbook for practicing Engineer.* Published by Penn well books.

NELSON, E.B, 1990. *Well cementing.* Published by Newes publication

ROBERT F.M. MITCHELL, 2006.Petroleum *engineering handbook.* Published by society of petroleum engineers.

WILLIAM .C. LYONS, GARY J. PLISGA, 2011.Standard Handbook of Petroleum and Natural gas Engineering. Published by Gulf professional publishing.

BIBLIOGRAPHY

ADAMS NEAL. J, 1985.Drilling *engineering*: *A complete well planning approach*. Published by Penn well books.

FRANK JAHN, MARK COOK, MARK GRAHAM, 2008.*Hydrocarbon exploration and production*. Published by Elsevier.

FERTL.W.H, 1981. *Abnormal formation pressure*. Published by Elsevier.

HUACAN TANG, MENGLAN DUAN, 2014.Offshore *operation facilities*. Published by gulf professional publishing.

JACK JEZOWSKI, JAN THULLIE, 2009.19th *European symposium on computer aided process engineering*. Published by Elsevier.

JAMAL AZAR.J, ROBELLO SAMUEL.G, 2007. *Drilling engineering*. Published by Penn well books.

JEANATTE PAXXSON, 1982.Casing *and cementing*. Published by petroleum extension service.

MARTIN RAYMOND, WILLIAM LEFFLER. L, 2006. *Oil and gas production in non-technical language*. Published by Penn well books.

PAL SKALLE, 2011.*Pressure control during oil well drilling*. Published by Pal skalle and ventures.

POLETTO F.B, MIRANDA F, 2004.Seismic *while drilling: Fundamentals of drill bit seismic for exploration*. Published by Elsevier.

RAMACHANDRAN .V.S, 1996.Concrete *Admixtures handbook*. Published by Noyes publication.

ROBERTSON J.O, CHILINGARIAN G.V, KUMAR.S, 1989.*Surface operations* in *petroleum production II*. Published by Elsevier.

APPENDIX

3.1.1 Determination of pore pressure at different depths.

Pore pressure using ppg and feet as the unit measure

Field units

P= Mud weight, ppg × 0.052 × true vertical depth (TVD), ft

Mud weight (ppg) = density of the fluid causing hydrostatic pressure.

True vertical depth (TVD) = depth at which hydrostatic pressure occurs.

P (psi) = hydrostatic pressure.

- The pore pressure at 4000ft

P= 8.9 × 0.052 × 4000

P= 1851.2 psi

- The pore pressure at 9000ft

P= 10.5 × 0.052× 9000

P= 4914 psi

- The pore pressure at 12000ft

P= 10.5 × 0.052× 12000

P= 6552 psi

4.2 Cementing calculations

Volume (bbls) = Capacity × depth

Capacity of Hole (bbl/ft) = $D^2 \times \frac{\pi}{4} \times 12 \div 1728 \div 5.615$

Capacity of Annulus (bbl/ft) == $D^2 - d^2 \ \frac{\pi}{4} \times 12 \div 1728 \div 5.615$

- The Volume of cement slurry required using an open hole excess of 30%.

Capacity of 13 3/8" and 9 5/8" annulus = $(12.347^2 - 9.625^2) \frac{\pi}{4} \times 12 \div 1728 \div 5.615$

$$= \frac{(12.347^2 - 9.625^2)}{1029.4}$$

$$= \frac{59.807}{1029.4} = 0.0581 \text{bbl/ft}$$

Volume of the cement in the casing = 0.0581 bbl/ft × 1000 ft

=58 bbls

Capacity of 12 ¼" hole and 9 5/8" = $(12.25^2 - 9.625^2)\ \frac{\pi}{4} \times 12 \div 1728 \div 5.615$

$$= \frac{(12.25^2 - 9.625^2)}{1029.4}$$

$$= \frac{57.422}{1029.4}$$

= 0.0558 bbl/ft

Volume of the cement in open hole = 0.0558 bbl/ft × 6000 ft

= 334.8 bbls

= 335 bbls

Because the open hole had excess of 30%

1.30 × 335 = 435.27 bbls

= 435 bbls

Capacity of the shoe track = $8.535^2 \times \frac{\pi}{4} \times 12 \div 1728 \div 5.615$

$$= \frac{8.535^2}{1029.4}$$

= 0.07076 bbl/ft

= 0.0708 bbl/ft

Volume of the cement in shoe track = 0.0708 bbl/ft × 80 ft

= 5.66 bbls

Total volume of the cement slurry = volume of the cement in the casing + volume of the cement in the open hole + volume of cement in the shoe track.

Total volume of the cement slurry = 58 + 435 + 5.66

= 498.66 bbls

= 499 bbls

- The volume of mud required to displace the cement.

Capacity of the shoe track = $8.535^2 \times \frac{\pi}{4} \times 12 \div 1728 \div 5.615$

$= \frac{8.535^2}{1029.4}$

= 0.07076 bbl/ft

= 0.0708 bbl/ft

Height of the mud = total depth – shoe track depth

Height of mud = 10,000 – 80 = 10,000 ft – 80 ft = 9920ft

= 0.0708 bbl/ft × 9920 ft = 702.3 bbls

=702.3 bbls

- Differential pressure at the float collar before the slurry starts to set.

Pressure= Mud weight, ppg × 0.052 × true vertical depth (TVD), ft

Pressure of mud $_{(casing)}$ = 11 × 0.052 × 9920 = 5,674.2 psi

Pressure of cement $_{(shoe\ track)}$ = 12.5 × 0.052 × 80 = 52 psi

= 5674.2 + 52 = 5726.2psi

Pressure of mud $_{(annulus)}$ = 11 × 0.052 × 2138 = 1,222.95 psi

Pressure $_{(spacer)}$ = 9 × 0.052 × 862 = 403.42 psi

Pressure of cement $_{(casing)}$ = 12.5 × 0.052 × 7000 = 4550 psi

= 1,222.95 + 403.42 + 4550 = 6176.4 psi

Differential pressure at the float collar = 6176.4 - 5726.2 =450.4 psi

= 450 psi

6.2 Drill collar calculation

Calculation of the number of the drill collar required in the BHA to provide a weight of 21,000 lb to drill the final section of the well.

<u>Assumptions</u>

- BHA comprises of drill collars
- The well is at the angle of 10 degree
- 6 3/4" ID 3" (Williams and Gary, 2011)

- A drill collar is approximately 30ft long

Weight of drill collar (lb/ft) $= 2.67\,(OD^2 - ID^2)$

$= 2.67\,(6.75^2 - 3^2)$

$= 97.62$lb/ft

Buoyancy factor = [1-0.0153 (mud weight)]

= [1-0.0153(11)]

= 0.832

Weight of the drill collar considering the height and the angle

$$W = \frac{21000}{\cos 10^\circ}$$

= 21324 lb

Required weight in air $= \frac{\text{weight}}{\text{Buoyancy factor}}$

$= \frac{21324 \text{ lb}}{0.832}$

=25630 lb

Total length of the drill collar $= \frac{\text{Weight in air}}{\text{Weight of drill collar}}$

$= \frac{25630 \text{ lb}}{97.62 \text{ lb/ft}}$

= 262.55 ft

Number of drill collar $= \frac{\text{Total length of drill collar}}{\text{Approximate length of drill collar}}$

$= \frac{262.55}{30}$

= 9 drill collars

Average drilling rate over the entire well

1 m = 3.2808 ft

6.25 m = 6.25 × 3.2808 = 20.505 ft/hr

Number of ft/day = 20.505 ft × 24 hrs = 492.12 ft/day

Total days to drill 1200ft $= \frac{12000}{492.12} = 24.38 = 24$ days